HARLEY-DAVIDSON®
— LORE —
SHOVELHEAD TO TWIN CAM 88

NINETEEN SIXTY-SIX  THROUGH PRESENT

HARLEY-DAVIDSON® — LORE —

SHOVELHEAD TO TWIN CAM 88

NINETEEN SIXTY-SIX  THROUGH PRESENT

FOREWORD: DR. MARTIN JACK ROSENBLUM, PREFACE AND TEXT: HERBERT WAGNER, DESIGN: VSA PARTNERS, CHICAGO

CHRONICLE BOOKS
SAN FRANCISCO

Copyright © 2000 H-D.
All rights reserved.

Published by Chronicle Books under license from The Harley-Davidson Motor Company. No part of this book may be reproduced in any form without written permission from the publisher.

Library of Congress Cataloging-in-Publication Data available.

ISBN 0-8118-2533-7

Printed in Hong Kong.

Harley, Harley-Davidson, Harley Owners Group, *The Enthusiast*, Fat Boy, and Electra Glide are among the trademarks or registered trademarks of H-D Michigan, Inc.

Distributed in Canada by Raincoast Books
9050 Shaughnessy Street
Vancouver, British Columbia
V6P 6E5

10 9 8 7 6 5 4 3 2 1

Chronicle Books LLC
85 Second Street, Sixth Floor
San Francisco, CA 94105
www.chroniclebooks.com

Designed by: Dana Arnett, Ken Fox, Michael Petersen, and Jason Jones of VSA Partners, Chicago.

Special thanks: Tom Bolfert, Mindy Christman, Scott Heaton, Bill Jackson, Joyce Muffoletto, Dr. Martin Jack Rosenblum, and Cheri Van Den Kieboom, of the Harley-Davidson Motor Company.

Harley-Davidson does not recommend certain riding practices or nonstandard equipment that might appear in any of the photographs featured in this book.

Personal statements are the sole opinion of the author and are not endorsed by Harley-Davidson Motor Company.

FOREWORD: LEGITIMATE FACT AND REAL LEGEND — DR. MARTIN JACK ROSENBLUM

RIDING A HARLEY-DAVIDSON MOTORCYCLE CHANGES YOU. ON THE BIKE, YOUR ENTIRE BEING IS AN INTOXICATING BLEND OF SENSATIONS. YOU ROUND A CURVE AT HALF THROTTLE, PIPES RAPPING, AND LEAN INTO THE TURN, BALANCING ON A DELICATE EDGE. AS YOU CUT THE APEX AND BEGIN TO ACCELERATE, THE WIND WHIPS AROUND YOU AND THE THRUST PRESSES YOUR BODY AGAINST THE BIKE. GRACE AND POWER BANISH EVERYDAY CARES. YOUR BODY RELAXES. FOR THE MOMENT, YOU ARE THE MOTORCYCLE, UNFETTERED AND FREE.

    BUT A HARLEY IS MORE THAN SIMPLY MOMENTARY PHYSICAL PLEASURE: HISTORY, IMAGE, AND FOLKLORE ARE VITAL PARTS OF THE ALLURE. EQUALLY IMPORTANT ARE THE BIKE'S THOUSANDS OF DEVOTED FANS, THAT SPECIAL COMMUNITY WHO UNDERSTAND AND LOVE THESE WONDERFUL MACHINES. HARLEY RIDERS TRAVEL HUNDREDS OF MILES FOR THE CHANCE TO MINGLE WITH OTHERS WHO SHARE THEIR ENTHUSIASMS. THEY PAY HOMAGE TO HARLEY BECAUSE THE

BIKE PROMISES — AND DELIVERS — BOTH PHYSICAL JOY AND SPIRITUAL ENERGY. IT IS NOT SOLELY THE GLINT OF CHROME, THE SWEEP OF A FENDER, THE RUMBLE OF POWER; RATHER, THE ESSENCE OF HARLEY OWNERSHIP IS SPIRIT. POWERFUL AND IMPOSING, HARLEY-DAVIDSON MOTORCYCLES EXEMPLIFY THE AMERICAN CYCLING EXPERIENCE, THE IMAGE OF THE SOLITARY TRAVELER ETCHED AGAINST A VAST LANDSCAPE, A SELF-RELIANT AND RESOURCEFUL PALADIN UNSWAYED BY FEAR AND INDECISION. WHEN A HARLEY TWIN RUMBLES TO LIFE BENEATH YOU, THE PULSE REACHES DEEP INTO YOUR BEING, CONVEYING THE PROMISE OF ESSENTIAL KNOWLEDGE AND PERFECT, INSPIRED LIVING.

AT ONE WITH OUR MOTORCYCLES, WE CRUISE THE OPEN HIGHWAYS, ALIVE TO POSSIBILITIES. IT IS THE BIKE, ALL HARD STEEL AND SUPPLE LEATHER, CONTROLLED POWER AND SPEED, THAT TAKES US TO A QUIET SPIRITUAL WORLD WHERE WE CAN CREATE A VISION OF WHO WE ARE — OR MIGHT BECOME. DR. MARTIN JACK ROSENBLUM, HARLEY-DAVIDSON HISTORIAN

PREFACE: *VOICES FROM THE PAST AND PRESENT—HERBERT WAGNER*

THE FIRST VOLUME OF *HARLEY-DAVIDSON LORE* TRACED THE HISTORY OF THE HARLEY-DAVIDSON MOTORCYCLE COMPANY FROM ITS BEGINNINGS IN 1903 TO 1965. DURING THOSE YEARS, THE COMPANY ESTABLISHED ITSELF EVEN AS THE NATION PROSPERED AND GREW INTO ITS ROLE AS A WORLD LEADER.

THIS SECOND VOLUME BEGINS IN 1966, WHEN THE U.S. AND HARLEY-DAVIDSON WERE IN UPHEAVAL. IN THE EARLY SIXTIES, THE HARLEY AND DAVIDSON FAMILIES MAINTAINED THEIR OWNERSHIP OF THE COMPANY. BUT IN 1969, IN A CONTROVERSIAL MERGER, AMERICAN MACHINE AND FOUNDRY (AMF) TOOK CONTROL. THROUGHOUT THE FOLLOWING TUMULTUOUS YEARS, AMF DIRECTED COMPANY FORTUNES UNTIL 1981, WHEN SEVERAL KEY AMF AND HARLEY-DAVIDSON EXECUTIVES PURCHASED THE COMPANY.

THE BIKES THEMSELVES SEEMED TO MIRROR CHANGES INSIDE THE COMPANY. IN 1966, HARLEY-DAVIDSON MADE JUST THREE V-TWIN MODELS: THE ELECTRA GLIDE AND TWO SPORTSTERS. BUT IN THE MOVIES AND IN BOOKS THE COMPANY BECAME A HOUSEHOLD NAME, FOREVER LINKED TO THE IMAGE OF THE NOTORIOUS GANGS WHO RODE LOUD, THREATENING

HARLEY-DAVIDSONS WHEREVER AND WHENEVER THEY WANTED. IN A NATION REBELLING AGAINST MANY CHERISHED IDEALS, THE CHOPPER AND ITS BIKER OWNER WERE SUDDENLY THE RAGE. CAUGHT OFF-GUARD BY THE SUDDEN POPULARITY OF ITS BIKES, THE HARLEY-DAVIDSON COMPANY WAS BOTH REPELLED BY THE LURID ASSOCATIONS OF ITS PRODUCTS WITH THE OUTLAW GANGS AND TANTALIZED BY THE MARKETING POSSIBILITIES. THE 1971 SUPER GLIDE CAPITALIZED ON THE POPULAR CHOPPER LOOK BY COMBINING THE LIGHTWEIGHT SPORTSTER FRONT CHASSIS WITH THE RUNNING GEAR OF THE BIG TWIN TO CREATE A FUTURISTIC DESIGN. OVER TIME, THE COMPANY DEVELOPED SEVERAL PLEASING FACTORY CUSTOM MODELS CENTERED UPON THIS PROVEN SUPER GLIDE MOTORCYCLE.

AS THE COMPANY EMERGED FROM THE TRIALS OF THE 1970s, IT DEVELOPED THE EVOLUTION ENGINE IN 1984,

WHICH SET A NEW STANDARD FOR THE COMPANY'S TRADITIONAL AIR-COOLED, PUSHROD V-TWIN. THE EVO WAS CLEANER AND MORE RELIABLE THAN ITS PREDECESSORS. IT WAS EVENTUALLY FOLLOWED BY THE TWIN CAM 88, A FURTHER IMPROVEMENT ON THE KNUCKLEHEAD AND PROOF POSITIVE THAT HARLEY-DAVIDSON REMEMBERED AND CHERISHED ITS OWN RICH HISTORY. TODAY, HARLEYS ARE RIDDEN BY PEOPLE FROM EVERY SOCIAL CLASS. NO LONGER AN OUTLAW ICON, IT IS A CHERISHED HOUSEHOLD NAME.

HARLEY-DAVIDSON LORE, NINETEEN SIXTY-SIX THROUGH PRESENT PAYS HOMAGE TO THE RECENT YEARS AND TO THE COMPANY'S INSPIRED DESIGNS. THE BOOK ALSO LOOKS FORWARD — WITH THE MOTOR COMPANY ITSELF — TO MOTORCYCLES THAT WILL CONTINUE TO CAPTURE IN ONE PLEASING PACKAGE BOTH SOPHISTICATED ENGINEERING AND MORE THAN FIFTY YEARS OF MOTORCYCLING LEGEND. HERBERT WAGNER, AUTHOR/SCHOLAR

THE SHOVELHEAD ERA—1966 THROUGH 1984

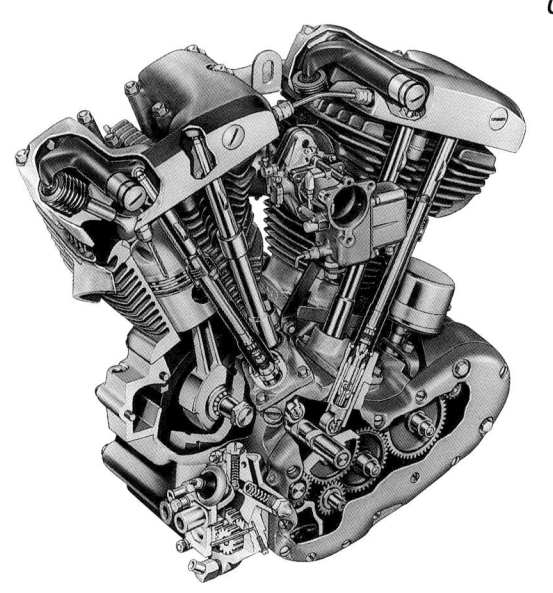

001

003

004

005

006

007

008

009

010

011

012

013

015

016

017

018

019

020

021

CERTIFICADO N° 1

quien queda inscrito en el registro respectivo como piloto de motocicleta, y por lo tanto sujeto al reglamento y disposiciones vigentes de la Inspección del ramo.

Nadie puede manejar una motocicleta sin poseer el certificado de competencia. Este certificado se obtendrá solicitándolo por escrito a la Inspección del ramo.

Las personas que pierdan sus certificados sufrirán una multa de Lp. 2 por primera vez y de Lp. 5 por segunda.

Lima, 3 de ABR 1922 de 192...

EL JEFE DE LA SECCIÓN

V.° B.°
EL INSPECTOR DEL RAMO

La Inspección de Rodaje, en conformidad con el Reglamento, expide el presente certificado a Enrique Santiago Lacerna

023

024

025

026

027

028

029

030

031

032

033

034

035

036

037

038

039

040

041

042

043

044

045

046

047

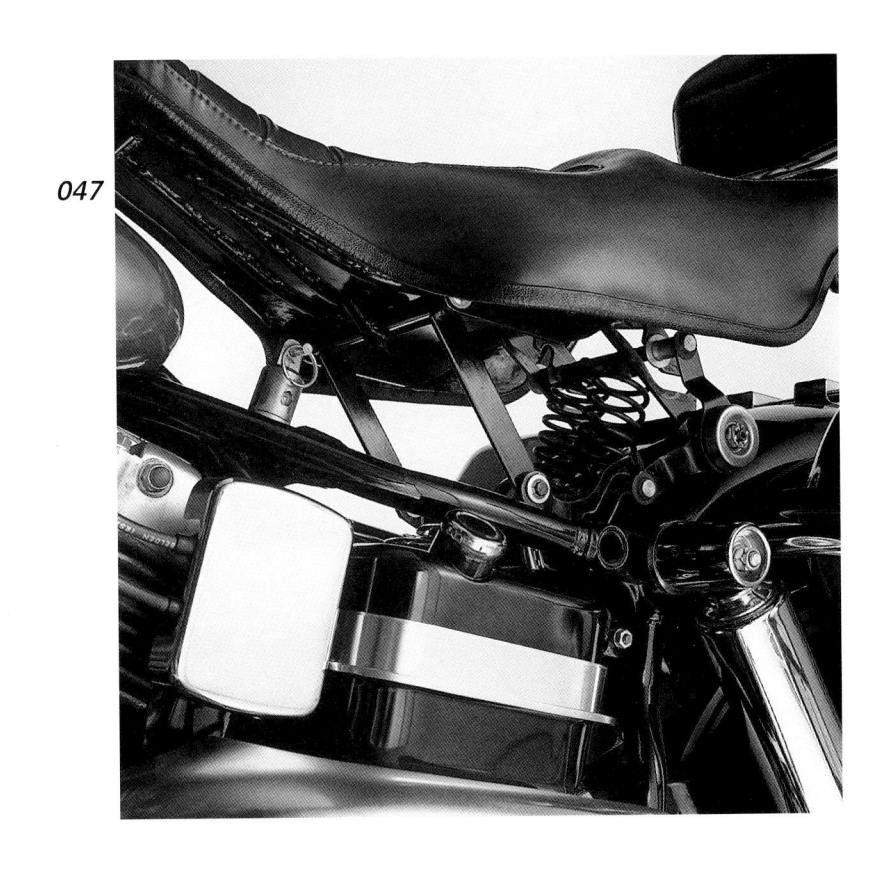

048

049

050

051

052

053

054

055

056

057

058

059

060

061

062

063

064

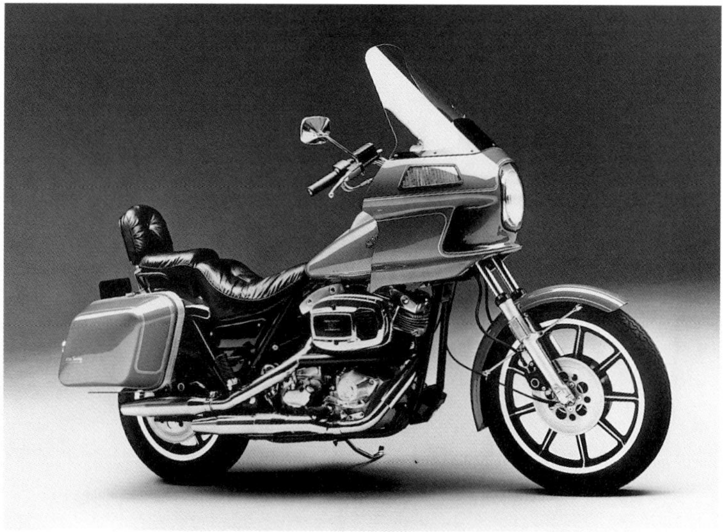

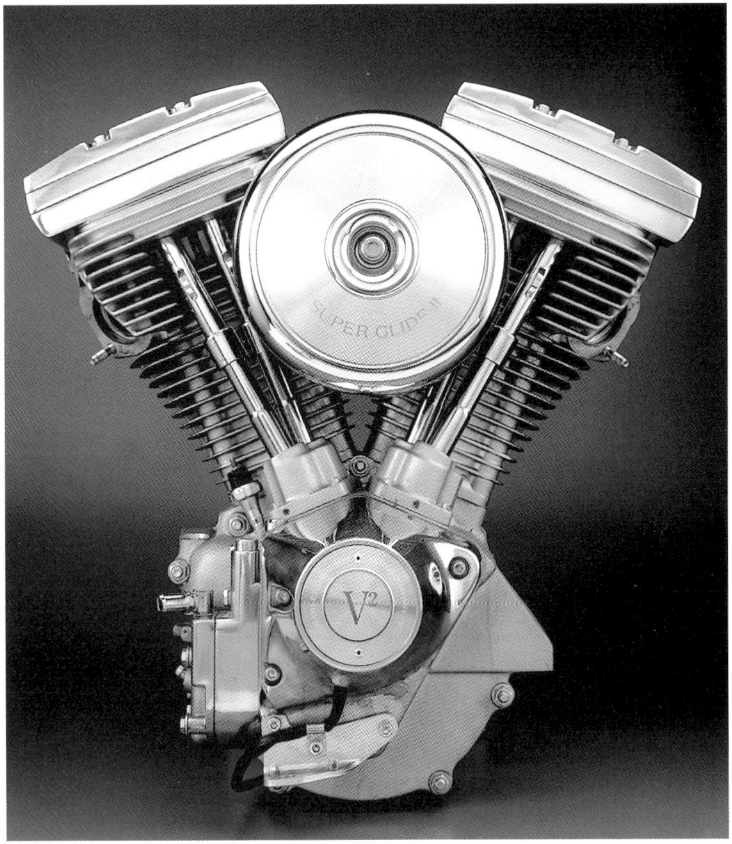
SUPER GLIDE II
V²

066

067

068

85

THE EVOLUTION ERA — 1985 THROUGH 1999

069

070

071

072

073

074

075

076

077

078

079

080

081

082

083

084

THE HOLY RANGER
HARLEY-DAVIDSON POEMS
MARTIN JACK ROSENBLUM

085

086

087

088

089

090

091

092

093

094

095

096

097

098

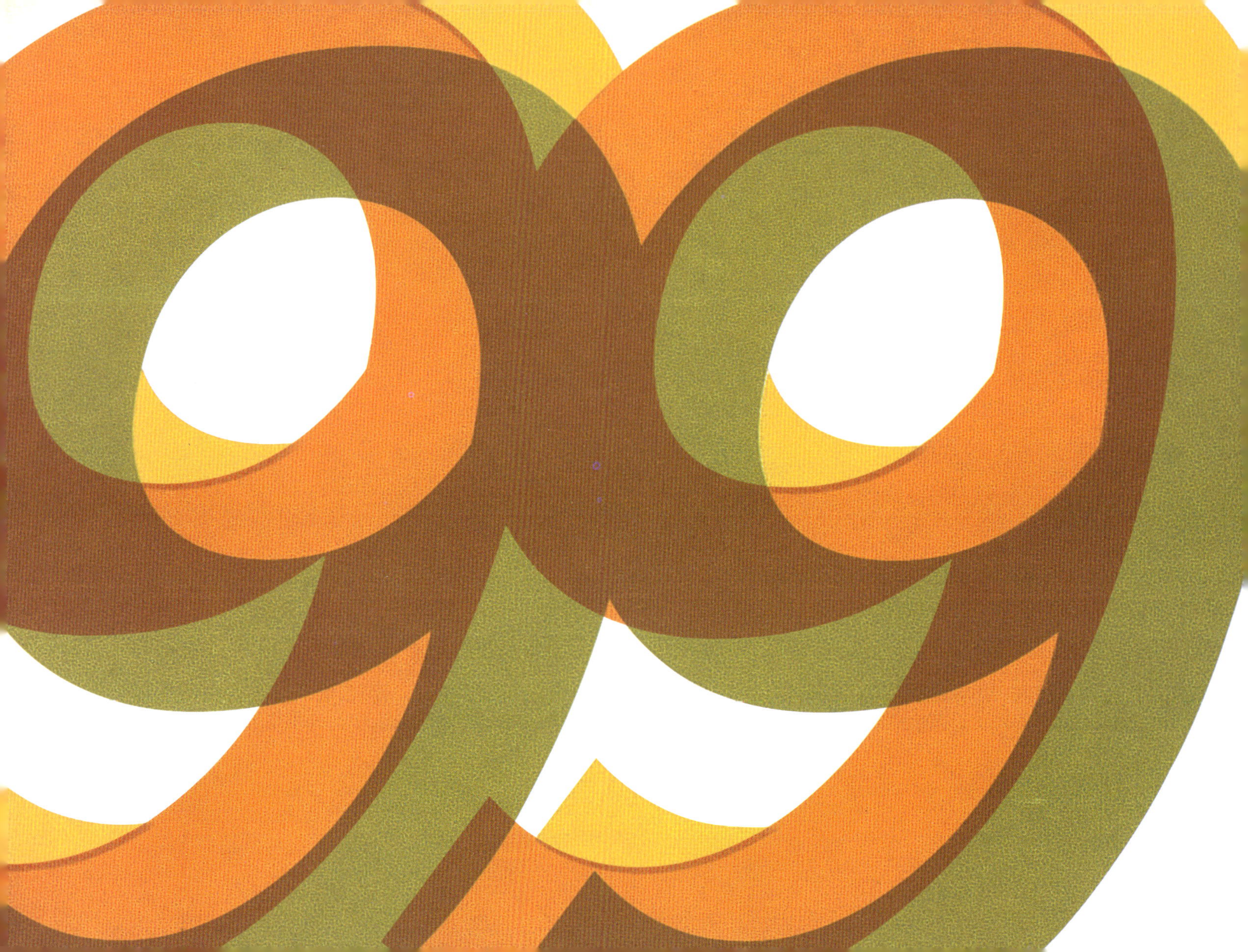
99

THE TWIN CAM 88 ERA — 1999 THROUGH 2000

099

100

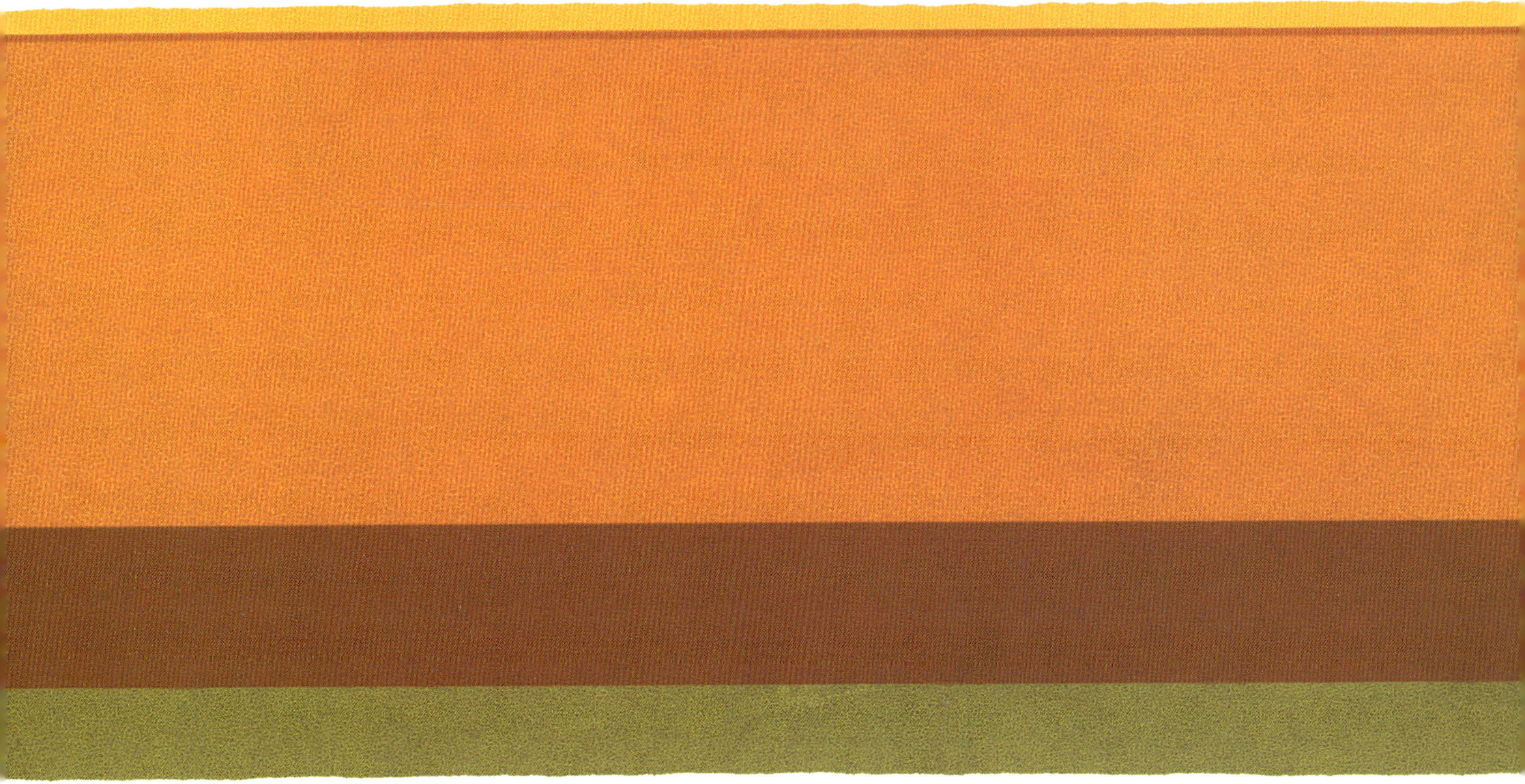

INDEX: ADVENTURES ON AN IRON STEED

The following photographs were reproduced from originals at the Harley-Davidson Archives in Milwaukee and from other sources. The captions are based upon period publications, advertising literature, personal recollections of riders and Harley-Davidson employees, and the author's own experience. They are a tribute to the world of Harley-Davidson motorcycling.

001 (1966) Some thought that the new Shovelhead Big Twin engine of 1966 looked more like the original 1936 61 OHV Knucklehead than the intervening Panhead engine. This was due to the hand of Charlie Featherly, who designed the heads of the Knucklehead, Sportster, and Shovelhead engines. The smooth cam cover was also a nod to the original 61 OHV. This loyalty to older engine types resulted in severe criticism of Harley-Davidson during the Japanese superbike era, but ultimately proved to be inspired thinking.

002 (1966) A growing concern for Harley-Davidson during the late 1960s was a solitary Big Twin offering in the Electra Glide. While company officials thought that the Electra Glide (right), two Sportsters (Model XLH on left), and the Italian-made lightweights covered the full market spectrum, some riders thought otherwise and were "chopping" Electra Glides and older Big Twins into lighter mounts. The Motor Company initially resisted this trend because of its association with the so-called outlaw biker.

003 (1966) This front fender style first appeared on the 1949 Hydra-Glide and with little modification has remained one of the Motor Company's longest-running parts. It served faithfully on the Duo-Glide (1958–1964) and presently does duty on the Electra Glide, Road King, and Road Glide. It's also used on the Heritage Softail line. Here the fender is seen on a 1966 Model FLH Electra Glide.

004 (1966) Unlike the single Electra Glide model, the Sportster came in two distinct forms. Here we see the highway-oriented XLH (left) and the dual-purpose XLCH. While the stripped-down, magneto-equipped, and hotter "CH" was intended to serve as a big-bore off-road competition bike, more often it was used to humble British-built iron on the street.

005 (1966) The Shovelhead was the last Big Twin to utilize a hemispherical combustion chamber. This configuration was first used on the 21 OHV Single of 1926; and later it appeared on the Knucklehead, Panhead, and Shovelhead V-Twins. The Shovelhead's new "power pac" heads gave it a reported 10 percent horsepower advantage over the Panhead engine and upped output to 65 ponies. This assured the Electra Glide's title as "King of the Highway."

006 (1966) The Shovelhead received restyled footboards that the factory described as having a "futuristic" look. This was interesting in light of the fact that footboards (sometimes called floorboards) were first used on the Harley-Davidson in 1914 and were one of the last links to a distinctly American style of motorcycle. After Indian's demise in 1953, the Harley-Davidson Electra Glide and Servi-Car were the only motorcycles to use footboards until recent years when several manufacturers have copied this practical item.

007 (1967) From its performance-inspired origin in 1936, H-D's overhead-valve Big Twin had grown by the 1960s into the Electra Glide. On this white-walled, heavy-weight touring bike a rider could pack all the comforts of home into spacious fiberglass saddlebags, smile bug-free behind a windshield, and offer the passenger a comfortable ride on the Buddy Seat. While this plush pavement-eater appealed to a whole generation of riders, the need for a sport-oriented Big Twin would ultimately push Harley-Davidson towards a new type of motorcycle — the factory custom.

008 (1967) Although the automobile nudged the horse into extinction as a means of basic transportation, the motorcycle remains true to the heritage of its equine ancestor. The riding position, wind-in-the-face exposure to the elements, and raw excitement far better replicate the characteristics of the horse than does the insulated interior of an automobile.

009 (1967) In the pre-AMF years of the late 1960s, Harley-Davidson motorcycles were still largely hand-assembled on the third floor of the Juneau Avenue factory using methods not much different from those of the 1930s. Another link with tradition was the kick-start — still standard equipment on this push-button start 1967 Electra Glide. But changing times were evident in a gas tank badge, which is similar to the new corporate logo adopted in 1964 to cover both the motorcycle line and the Golf Car.

010 (1967) This photo is so pleasantly evocative that I had to include it. Nice, normal, young people unaffected by the churning social currents of the 1960s meet on a quiet country road in the Dairy State. She's out with the horses on the farm when he comes along on a spanking new Electra Glide — the eternal story of boy meets girl. Who cares that he's a factory guy chosen because he can handle a Big Twin and she's a professional model and the scene is staged. It looks real to me.

011 (1969) This looks like business as usual and another racing victory for Harley-Davidson as president William H. Davidson congratulates Cal Rayborn at Indianapolis. Racing director Dick O'Brien stands behind them. But things at Harley-Davidson were anything but normal. In January, American Machine and Foundry Co. (AMF) had acquired Harley-Davidson and brought family ownership at the Motor Company to an end. That same year the American Motorcycle Association (AMA) dropped the "equivalency rule" that gave H-D's KR side-valve racing engine a 250 cc displacement advantage over overhead-valve motors. Challenging times lay ahead for Harley-Davidson.

012 (1969) Made famous by the movie *Easy Rider*, the Captain America chopper is probably the single best known motorcycle ever built. The fate of the original bikes used in the movie is sad to tell. Two each of the Captain America and Billy Bike choppers were built using retired police FLs. Two of these were blown up during the movie's final scene. The remaining two choppers were stolen shortly after the movie was finished and never seen again. Created by James Beck, this replica version was recently acquired by the Harley-Davidson Archives.

013/014 (1969) Mike Smith catches this great helmet shot of his brother Paul as he turns to speak to his son Miles on the porch steps. Seen here on his newly acquired 1951 WLA "beater," designer/artist Paul Smith's custom "skid lids" were some of the best seen around Milwaukee in the late 1960s. His most interesting creations were those with a face painted on the helmet's rear surface. With their big eyeballs, toothy grins, shark faces, and flames, these wild helmets were a big hit with fellow riders. (Copyright Notice: © Paul Smith 2000/Photo Credit: Mike Smith)

015 (1970) When the AMA abruptly dropped the 750 cc side-valve / 500 cc overhead-valve equivalency rule, Milwaukee turned to the iron barrel Sportster as a stop-gap racing engine. Despite its roadster origins and problems with overheating, the de-stroked iron XR-750 in a KR-style frame with Girling shocks and Ceriani front fork did fairly well in the hands of Cal Rayborn, Mark Brelsford, and Mert Lawwill during the two seasons it was campaigned. Because many iron barrel XR-750s were dismantled, surviving examples are highly prized by collectors.

016 (1970) AMF had scarcely unpacked their bags in Milwaukee when Cal Rayborn piloted a Sportster engine–powered streamliner to a new motorcycle speed record at Bonneville Salt Flats. Lying nearly prone and looking out of the two side windows (which may have contributed to two previous crashes), Rayborn flew down the 10-mile course at an average 2-way speed of 265.492 mph. Legendary racer Joe Petrali, who had established a speed record in 1937 on a Knucklehead, witnessed Cal's ride. Today the red, white, and blue "Number 1" emblazoned streamliner resides at the Indianapolis Motor Speedway Museum.

017 (1970) In 1969, Paul Smith purchased this 1951 WLA 45 for $125 (less twenty bucks for repairing a broken shifter fork). Not a radical chopper guy, Smith took a beater "bob job" to the next level. He rebuilt the engine and tranny, finding original WLA parts at a local shop that sold accessories for vintage Harley-Davidson motorcycles. With a new front tire, fresh paint job, Bates solo seat, and his own custom silk-screen tank decals that were inspired by early Harley lettering, Smith built this streamlined 45 for a few hundred bucks. (Copyright Notice: © Paul Smith 2000/Photo Credit: James Middleton)

018 (1971) This is the 1971 Super Glide in all its chopper-esque glory. When introduced, the Super Glide was considered quite daring around Harley-Davidson. The company was leery of the chopper image and any hint of extremism. To tone down the chopper look, a fiberglass tailpiece was conceived and constructed; otherwise the Super Glide was created out of items from the Big Twin and Sportster parts bins. Over the years, the Super Glide has changed a great deal, but the original remains distinctive, unique, and ageless — Willie G.'s first factory custom.

019 (1971) Mark Brelsford was one of H-D's bright spots during the troubled time of the iron barrel XR-750. Brelsford came on strong in 1971 by winning the Loudon 100 — the only AMA national road race ever won by the iron barrel XR. Here Brelsford takes first place at the Ascot TT in 1972 on a 900 cc XLR. That year he won the Grand National Championship and captured the No. 1 plate for H-D.

020 (1971) The 1971 Super Glide first appeared in a kind of sneaky way. After a Sportster front end was grafted to a stripped FLH chassis, the seat/fender unit was fabricated from wood. The seat area was covered with fabric. From a few feet away, it looked stock. A photo of this mock-up was published in the 1969 *Motorcycle Sport Book*, which slyly described the bike as the work of an "enthusiast" (Willie G. — shown here in later years) operating on "distant shores" (Lake Michigan). But observers saw through the deception and clamored for the new-style machine.

021 (1972) The early 1970s saw long-distance motorcycle touring becoming more popular. In 1972, *The Enthusiast* ran a two-part article by Michael Sumner, who, during the previous year, had ridden from New York City to South America. During part of the trip Sumner had a camera tripod rigged to the front fork of his Electra Glide. He took this self-portrait that shows road conditions encountered south of the border.

022 (1972)　　An interesting photograph that Sumner sent back to *Enthusiast* editor Tom Bolfert was of Peruvian H-D dealer Enrique Pontillo's 1922 motorcycle license. This historic document was the first motorcycle license ever issued in that South American country. That same year Pontillo opened an H-D dealership. Fifty years later he was still in business, although high import taxes had diminished his client base to just one: the motorcycle guard of the Peruvian president.

023 (1972)　　In spite of first appearances, this isn't your typical male drag racer in action. Rather it's Mrs. Mary Baisley of Portland, Oregon, popping a wheelie as she comes out of the hole at Phoenix, Arizona. Mrs. Baisley had been an arts and crafts instructor when she caught the motorcycle bug from her husband. In the Baisley family everyone pitched in. Father and husband Wes built the bikes that his wife Mary and son Dan raced. Mary's best quarter-mile time in 1972 was 12.97 seconds at 105.67 mph on her Sportster-powered dragster.

024 (1972)　　Desert scrambles and enduros were popular sports during the early 1970s. Harley-Davidson responded to this trend by marketing the "Baja." This Italian-made 100 cc two-stroke lightweight was popular among younger riders. Here fifteen-year-old Bruce Bornhurst pops a wheelie on his Baja during the Calico Ghost Run, where over 2,100 bikes and riders of all ages ran a 162-mile course through the California desert.

025 (1972)　　In 1967, Evel Knievel was injured while jumping thirteen Mack trucks at Caesar's Palace. After lying in a coma for twenty-nine days, the world's greatest stunt man conceived his ultimate dream: jump the Grand Canyon. Back on his feet, Evel began making plans. After park officials rescinded their initial go-ahead, Evel moved his epic jump to Idaho's Snake River Canyon. Here we see Evel's rocket-assisted X-1 Skycycle. This full-size pilotless model was used to test the trajectory and distance of the Snake River jump, which, when attempted in 1974, was not successful.

026 (1973)　　Calvin Lee Rayborn was one of the greats on the H-D factory racing team during this period. His first major win was the Carlsbad National in 1966. From there Rayborn's career skyrocketed with first-place finishes at the Daytona 200 in 1968 and 1969. While Cal never excelled on the dirt track, he was a top-notch road racer, regularly winning up to half of the AMA National Championship races. His peak came during the Anglo-American Challenge Cup in 1972, where he took three wins and three seconds and was named "Man of the Meeting" by the British.

027 (1973) What these Harley lightweights were doing in a mountainous snowfield and who the riders were is a mystery. When this photo appeared in the early 1970s (the bikes are older), Harley's lightweights were still competitive with Japanese bikes. But as the decade progressed this situation rapidly changed. The last H-D lightweight was the Italian-built two-stroke SX-250. It was discontinued in 1978, and there hasn't been a Harley lightweight since.

028 (1973) Three Los Diablos club members and a "tag along" are seen with their bikes. (Left to right) Odel Reed, H. "Happy" Jackson, Leonard Smith, and Norman "Bird" Robinson. These were all family men, with the youngest member being thirty-three years old at the time. With their Harley hats and well-equipped and perfectly maintained dressers, members of clubs like Los Diablos were more akin to the AMA riders of the 1950s than the newer chopper-inspired Harley riders of the late 1960s and early 1970s.

029 (1973) The new alloy XR-750 racer blasted onto the scene in 1972 with well-deserved fanfare. Based on the Sportster engine, the new XR included several innovations that eventually found their way into the V2 Evolution. These included iron-sleeved aluminum cylinders, one-piece flywheel and mainshafts, and long cylinder studs instead of head bolts. At the time of the alloy XR's introduction, the V-Twin was so out of favor that Harley-Davidson had to remind the public that there was not anything inherently wrong with the V-configured motorcycle engine. How times change.

030 (1973) During the KR model's last year on the track in 1969, Mert Lawwill stunned the competition and took the AMA National Championship. Lawwill later starred in the 1971 motorcycle documentary *On Any Sunday*. In the movie, this H-D team racer was tracked by a camera crew throughout an entire competition season. After retiring from active racing, Lawwill became a top XR tuner for Harley rider Chris Carr. Lawwill is seen here at the Indianapolis Fairground races.

031 (1974) During the expansion of the Indianapolis Motor Speedway Museum, H-D president William H. Davidson (center) was asked to find a suitable Harley-Davidson motorcycle to display there. In Chicago he found a rough 1930s racing chassis for which H-D employee Ralph Berndt donated a mint Peashooter engine. Jim Haubert then rebuilt the machine to the specs of the single-cylinder Harley that Joe Petrali campaigned in 1935 — the year Joe won all thirteen National Championship dirt track races. Museum owner Tony Hulman is sitting next to the bike. Can you guess who that third fellow is?

032 (1975) When Willie G. Davidson selected Milwaukee freelance designer/illustrator Paul Smith to create the 1976 Bicentennial "Liberty Edition" tank decals, a large oil-on-canvas painting was also commissioned. Reaching deep into the Motor Company's history and his own Harley riding experience, Smith resurrected the Bar & Shield and united it with his rendition of the American bald eagle. With upswept wings, powerful talons, and a sinister glance to its left, Smith surrounded his new Eagle on the Bar & Shield with significant Motor Company history in an illustrative multi-subject montage. Intended as a Bicentennial project only, the nature of Smith's Liberty Edition graphics hit the mark and have been with Harley-Davidson ever since. (Copyright Notice: © Paul Smith 2000/Photo Credit: Bob smith)

033 (1976) By the mid-1970s, the quality of AMF/Harley-Davidson motorcycles was widely perceived to have slipped. To make matters worse, Japanese factories were now turning out motorcycles in the "super" heavyweight class (over 750 cc). But while Milwaukee's market share shrank and a crisis was in the making, AMF/Harley-Davidson could still point to the higher resale value of its machines. Today this 1973 FX model would be worth considerably more than when new.

034 (1976) Seen here from 800 feet up, Utah's "Widowmaker" had never been topped until the spring of 1976 when Californian Jim True blasted over the crest of this 1500-foot high bump on the landscape. True's Sportster hybrid was built using XR, XLR, and XLCH parts. After reaching the top, True remarked, "It was freaky. There was absolute silence up there. I couldn't hear the crowd (12,000 strong), and no one was on top of the hill... it was like being in another world."

035 (1977) The XLCR-1000 "Cafe Racer" was Harley-Davidson's sexy reply to what guys were doing to British bikes in the late 1960s. Willie G. described the CR as a "dream project... between designer and prototype fabricator." Much of the inspiration came from the XR racer. The prototype was built in Jim Haubert's shop a few miles from Juneau Avenue. The CR was described as a "hairy-looking, good handling, good braking, high performance Sportster." It was so different in fact that most traditionalists shunned it, and this future collector's dream was phased out after a two-year production run.

036 (1977) After returning from Daytona Bike Week in 1976, Willie G. said, "the collage of all that machinery was filling me with ideas." These ideas translated into the FXS Low Rider. Because lead time was critically short for a 1977 introduction, engineering chief Jeff Bleustein gathered his people together one evening after work at a local pub. With red markers flashing, Willie G. and Louie Netz went through every Super Glide drawing that had to be changed to create the Low Rider. These assignments were then handed out to the engineers and draftsmen.

037 (1978) To celebrate H-D's 75th Anniversary, company chairman John Davidson and president Vaughn Beals planned seven rides by Motor Company executives from distant parts of the country to Milwaukee. When the rides were completed and company execs had tallied up over 37,000 miles, Harley watchers had new respect for the Milwaukee product and for those who ran the company. Here (left to right) Vaughn Beals and advertising and promotion manager Clyde Fessler visit the York final assembly plant with H-D district manager Bob Jerrahain after departing Lewiston, Maine on May 25.

038 (1978) Steve Lawson had two Harleys stolen before hitting upon a fool-proof security system. This took the form of Dobs, a 140-pound mutt of uncertain origin who not only held down Lawson's sidecar during hard right-hand turns but stood guard while his master was absent. According to Lawson, Dobs grew up in the 1958 Duo-Glide sidecar rig and didn't take kindly to strangers getting too close to it.

039/040 (1978) Dirt track racing is America's oldest form of organized motorcycle competition, originating on county fair horse tracks before 1905. None of the excitement had been lost in 1978 as the alloy XR-750 battled Yamaha's 750 cc vertical twin for dirt track supremacy. Throughout the year Jay Springsteen and Steve Ekland exchanged leadership positions in a point contest until "Springer" emerged victorious. This was Springsteen's third straight year as Grand National Champion. He was the first rider to accomplish this feat since Carroll Resweber. Here (top) we see H-D team members Jay Springsteen and Ted Boody and (bottom) Corky Keener riding XR-750s.

041 (1978) Harley-Davidson's big V-Twin has always been known for its heavy-duty clutch and stump pulling torque. However, enthusiasts sometimes take this a step too far, as in the tale of a man who pulled down a farm shed with his Panhead chopper while at a biker party in the 1970s. Here we see a police officer demonstrating the Big Twin's power by towing a motor home. Harley-Davidson does not recommend using V-Twin power in these ways.

042 (1978) During the 75th Anniversary ride, Dick Reiter, parts and accessories director, teamed up with vice president of engineering Jeff Bleustein. In those pre–belt-drive days, chain adjustment and lubrication were necessary maintenance on the 1,733 miles they covered after leaving Schenectady, New York. Here (according to the photo's original caption), "Dick Reiter is caught by the hidden camera using something other than 'Genuine' Harley chain spray."

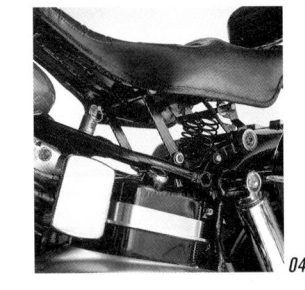

043/044 (1978) As a show of loyalty to the Milwaukee brand of motorcycle, Harley-Davidson dealers in Northern California hosted the first annual Redwood Run in June of 1978. The $5 participant fee covered two meals, a campsite, souvenir pin, and all the fun you could handle including a Harley-only ride from San Francisco to Garberville along picturesque U.S. Highway 1. Here some of the 1,400 Harleys that attended arrive at the lodge and campground, while other riders line up for the tasty barbecue chicken dinner.

045 (1978) To celebrate Harley-Davidson's 75 years of American motorcycling, Milwaukee authorized a series of large (3 ft. x 4 ft.) color posters. The most interesting was this historical collage of bikes and events including the belt-drive Silent Gray Fellow and Grand National Racing Champion Jay Springsteen. Today, this poster that originally sold for $2 is a highly regarded collectors' item.

046 (1978) If you were a motorcycle buff during the 1970s, you might remember seeing bikers R.L. "Cotton" and Mabel Steelman on the road. The retired Texas couple had taken several long trips on their Electra Glide before this photo appeared in Harley-Davidson's *Enthusiast* magazine. Already having toured the Pacific Northwest, the Southwest desert, and middle America, they were planning a new excursion to the East Coast.

047 (1978) Dubbed the Hi-Lo Comfort Flex Seat, this curiosity appeared as an option beginning in 1978. It had been brought to H-D by an outside supplier who spent a great deal of time and money developing it. While this seat suspension system actually worked quite well, it was complex, ugly as sin, and called for a long inseam to reach the ground. Riders snubbed it, and the Hi-Lo Comfort Flex Seat died a natural death in 1980. There were no mourners.

048 (1978) In 1976, Harley-Davidson began holding its own motorcycle exhibit in a hotel ballroom during Daytona Bike Week. To kick off the 75th "Diamond Anniversary" festivities in the spring of 1978, a display of new, vintage, and custom Harley-Davidson motorcycles were enjoyed by over 40,000 visitors. Here brothers Willie G. and John Davidson shake hands after cutting the ribbon that officially opened the Daytona Hilton exhibit.

049 / 050 (1978) During the 75th Anniversary ride, two young guys were given the "Easy Rider" route through the deep South from Texas to Kentucky. But sales promotion manager Tom Bolfert and sales office manager Tom Platz came up with a sure strategy for survival. They picked factory license plates 41HE and 41HA and rode so their plates read "HE-HA" to those who followed them. Maybe that's why they experienced nothing but old-fashioned Southern hospitality along the way. In Memphis they got "the royal treatment" when touring Presley's Graceland Mansion and were made "Honorary Deputy Sheriffs" of Shelby County, Tennessee.

051 (1978) For Harley-Davidson's 75th "Diamond" Anniversary, AMF/Harley-Davidson refurbished one of the oldest bikes in the factory collection, then proudly posed it with the new models. Here a 75th Anniversary Sportster is seen with the early single rising out of the mists of time. Like the first Harleys, the Anniversary Sportster was finished in black, but unlike its predecessor, the 75th had gold-enhanced cast wheels, gold hand-applied pin-striping, gold-lettered air cleaner insert, and gold-accented tank graphics borrowed from H-D's past.

052 (late 1970s) AMF/Harley-Davidson was considering adding this Italian-made 350 cc V-Twin to the line-up and tested a prototype machine. Harley engineers reported back that the bike was slow, vibrated badly, and was mechanically defective. In spite of the AMF/Harley-Davidson emblems seen here on a prototype, it never became a production model.

053 (1979) Taking a cue from their success with the 1977 Low Rider, H-D decided to give the world a factory customized Sportster. The new 1979 XLS Roadster used the XLCR Cafe Racer's triangulated frame and rectangular tube swing-arm, double front disc brakes, relocated shocks, siamese exhaust, and side covers. Riders were underwhelmed and most didn't take kindly to this messing around with the traditional Sportster look. As a result Sportster sales slumped and the XLS soon saw another revamping. The lesson was clear — don't tamper with success.

054 (1980) This photo of a twin-belt Sturgis FXB left me wondering. See anything unusual? How about those disc wheels? As far as I can figure, the Sturgis was only offered with cast spoke wheels and not these babies with the cool looking socket head screws. Is this a styling mock-up for a Sturgis variation that was never released? Or was Willie G. starting to play around with disc wheels for a new model that in time evolved into the FXDG Disc Glide?

055 (1980)   This photo and the two that follow first appeared in the AMF/Harley-Davidson 1980 *Motorcycle Fashions and Accessories* catalog. Paging through my old copy, I noticed items I bought back then that are no longer available: the sheepwool-lined winter riding mitts ("still the top performer for cold weather riding"), the six-inch-wide "Super Body Belt" (that my dealer thought nobody would wear so he gave me a deal on one that I still use), and one of H-D's greatest all-time T-shirts: "And on the 8th Day GOD Created Harley-Davidson." There was no claim that God created AMF.

056 (1980)   Another shot that ran in the 1980 catalog captures the feel of road camaraderie as two bikers pass on the highway with the (simulated) headlights of an automobile in the background. Note the special manufacturer's license plate issued to H-D by the Wisconsin Department of Motor Vehicles. For many years these plates have been issued with two-digit numbers in the low 40s and a two-letter suffix. They are an easy way to identify Motor Company machines at rallies, on the road, and in photographs.

057 (1980)   With the studio lights cropped out and ignoring the fact that the plant is potted, this appears to be an ideal campsite right down to agreeable companions and the Wide Glide model. The only things missing are mosquitoes — or is that one perched on the traveling bag? No, it's the eagle logo of AMF/Harley-Davidson's short-lived line of Eagle's Nest Camping Equipment. In the photo this includes biker luggage, tent, sleeping bag, mess kit, and AMF/Harley-Davidson cooking grid.

058 (1980)   As long as motorcycles have existed riders have felt a strong urge to strap on a sleeping bag and go places. From the size of his pack and the rough terrain, Tom Akins and his Super Glide appear to be a long way from home in this 1980 photo.

059 (1980)   The FLT Tour Glide brought the Electra Glide back to life as a top touring mount. The blueprint in the background shows construction details of the FLT's rubber-mounted engine and "balanced" front end. But it doesn't reveal that the Tour Glide's steering geometry was inspired by H-D's Topper scooter. One day in the experimental department, they had a Topper up on a stand when Art Kauper noticed that no matter where they positioned the Topper's fork it stayed there. This led to discussion and ultimately to the "power steering" effect found on all of H-D's Touring bikes today.

060 (1983) This Harley Owners Group (HOG) rally in 1983 at Elkhart Lake, Wisconsin, was an early gathering of the organization that was established for riders of Harley-Davidson motorcycles. Announced during the winter of 1982–83, the club's membership reached 33,000 by year's end. Local chapters formed all over the country and soon the world. *Hog Tales*, the official publication of the Harley Owners Group, grew from a simple newsletter to today's slick full-color magazine. This idea to strengthen rider fellowship would succeed beyond the Motor Company's wildest dreams.

061 (1983) Although draped in the American flag, the Electra Glide police model had fallen upon hard times. Even so, a couple of unique features are visible on this 1983 FLHT. Note the air-shock seatpost designed for long tours in the saddle. Harley-Davidson's poor reliability during the AMF period had driven H-D's biggest law enforcement client — the California Highway Patrol — to buy Japanese bikes. In 1984 there was a total revamping of the H-D cop bike line, and that year the CHP bought 135 FXRP police motorcycles — their first Harley purchase in eleven years.

062 (1983) The quality of aftermarket hop-up parts has always been questionable. To address this problem, H-D introduced their line of Screamin' Eagle Performance Parts in 1983. These were backed by Milwaukee's long engineering experience and extensive experimental testing labs. They were designed to work correctly right out of the box. From the first camshaft and exhaust kits, the Screamin' Eagle line has expanded and today includes big-bore kits for the Twin Cam 88 that boost displacement to 95 cubic inches and give Harley's newest motor an eye-popping 100 pounds-feet of torque.

063 (1983) This artistic shot might be dubbed "Sportster Power." The 1983 XLX was an exercise in simplicity: flat black pipes, brushed cases, one color option, and a retail price of just $3,995. The XLX was a tribute to the stripped British-eating Sportsters of the early 1960s. The idea had come from Harley designer Louie Netz, who said, "We've made standard models, and… deluxe editions. I thought it was time we made a raw, gutsy motorcycle, a 'two-wheeled street rod.'" Willie G. added, "I like it because it's back to basics, where Harley-Davidson was in the first place…"

064 (1983) Erik Buell was project manager for one of the best road Harleys ever built: the FXRT Sport Glide. Based on the FXR "rubber glide" chassis, the RT was Harley's push for a more racy touring bike. The RT fairing was actually designed for the ill-fated Nova, with the large "rider ventilation" air intakes originally intended to cool the Nova's radiator. The demise of the FXRT was a testament to rider taste and not to any fault with a bike that keeps you dry in the rain and provides excellent wind protection. One that also has weather-proof saddlebags and weighs a hundred pounds less than the Electra Glide.

065 (1984) For years rumors circulated of a radically different Harley-Davidson engine under development. AMF/H-D sank millions into the OHC V-1100 and the liquid-cooled V-4 Porsche-Nova project. But when the "new" motor was unveiled in late 1983, Harley lovers sighed in relief. The Evolution motor was instantly recognizable as heir to the 61 OHV Knucklehead that Bill Harley breathed life into back in 1936. Alloy cylinders and other improvements vastly improved the old war-horse. Even the formerly hostile mainstream motorcycle press fell in love with the oil-tight, cooler-running, and more reliable Evo.

066 (1984) Here's a view of the concealed shocks on the Harley-Davidson Softail line. Frames of this type actually go back to 1901, but were forgotten for many years as the hard-tail or rigid frame became universal until the Model K in 1952. Yet the lines of the hard-tail were so clean and classic that H-D adopted the "Softail" frame in 1984 to recapture the older look. Beneath the frame, however, were cleverly hidden shocks to give this rigid frame look-alike a modern ride. The Softail was a hit with Harley fans, and soon became a best-selling model.

067 (1984) The concept of the XR-1000 was a great one: slap heads from the XR-750 racer on a pre-Evo Sporty bottom end, stick the resulting motor in a street chassis, and then splash racing director Dick O'Brien's name on the advertisements. Presto! Instant import killer! Alas, the XR-1000 didn't quite pan out. The bike was noisy, expensive, and the carbs were in the way. Most damning of all, it wasn't much faster than a hopped-up XLH. While the XR-1000 looked great in the ads — and still does — it vanished after two seasons.

068 (1985) As the Statue of Liberty approached its 100-year mark, a restoration fund-raiser was launched. H-D got involved when Willie G. and Vaughn Beals led two groups of riders across America to Washington, D.C. On September 22, some 700 Harleys rolled into the nation's Capitol where Beals handed over a $250,000 check for the statue's face lift. Behind the scenes, however, H-D was in serious financial trouble. Beals and Rich Teerlink barely managed to refinance the Motor Company in a cliff-hanging scenario. Perhaps Lady Liberty had given them luck.

069 (1986) Here we see another styling coup: the Heritage Softail. Once the FX Softail frame was in production, it was probably inevitable that Willie G. would do a 1950 FL restoration project. Most old Harley buffs agree that the look of the 1950 Panhead is neck-in-neck in sheer design beauty to that other all-time great, the 1936 Model EL Knucklehead. Except for the custom pipes and seat, the Heritage Softail would make even an expert do a double-take on this 1950 look-alike.

070 (1986)   Here's one of the first Buell motorcycles built — probably prototype #2. Erik Buell, formerly an engineer with Harley-Davidson, struck out on his own to build the best street-legal road racer in the world. Powered by H-D's XR-1000 engine, the rest of the RR1000 was largely Erik Buell's creation. With full bodywork, solo seat, and "limited steering lock," the RR was more racer than road bike. In 1987, Buell drummed up twenty-five orders from H-D dealers and production began. In 1993 Harley-Davidson Motor Company purchased 49 percent of Buell and an additional 49 percent in 1998. Today, Buell is the fastest-growing motorcycle brand in the world.

071 (1987)   Automobiles and motorcycles have been racing on the smooth sands near Daytona Beach, Florida, for nearly a hundred years. In 1937, the American Motorcycle Association sanctioned motorcycle races that were run over a combined road-and-beach course. In the decades after World War II, Daytona became a major biker gathering. Today, nearly half a million riders converge on the beachside community to meet other riders, share stories, and check out the bikes parked or riding along Main Street. Here riders cruise past the Daytona Hilton during Bike Week in 1987.

072 (1987)   Early in their history, Harley-Davidson owned their own trucks and even had a Milwaukee-built Stegeman. These trucks were later phased out and retired test riders recall how a breakdown away from the factory would result in a sidecar coming out with a tow rope to rescue them. All this changed when final motorcycle assembly was moved to Pennsylvania, and trucks were needed to carry components from Milwaukee to York. Since 1985, Harley-Davidson has owned their own truck fleet with the trailer rigs displaying Willie G.–designed graphics. Here two H-D semis are seen parked at a 1987 rally.

073 (1987)   All types of outlandish machines can be seen at motorcycle events, from radical choppers to customized full dressers. Some riders seem to have a fetish for extra lights and there are even contests for heavily accessorized bikes. With the extensive miles of travel ahead, there just might be a lifetime supply of lightbulbs in that trailer.

074 (1987)   Female motorcycle riders go back to the early part of the century. In Milwaukee, the first woman to register a bike was back in 1911. She was Crystal Haydel, office secretary to Walter Davidson. Back then women wore long dresses while riding. Fashions have changed today but not the excitement of owning a Harley-Davidson motorcycle. Here we see an enthusiast with her customized bike at a Daytona bike show in 1987. Willie G. is standing to the left of the bike — an FXSTC Softail Custom with "Harley Fox" graphics.

075 (1987)   When President Ronald Reagan arrived by presidential helicopter at H-D's final assembly plant at York, Pennsylvania, he was greeted by a throng of flag-waving employees demonstrating renewed confidence in Harley-Davidson. Since the time of the 1981 buy-back from AMF, Harley executives had turned a company facing extinction at the hands of Japanese competition into a once-again proud American success story.

076 (1987)   The president was at York by invitation of H-D chief executive officer Vaughn Beals, who had recently asked the International Trade Commission to drop the remaining import tariff restrictions on heavyweight Japanese motorcycles. While President Reagan's patriotic speech was widely covered in the national press, his words when starting up a new 30th Anniversary model Sportster are better remembered around the York plant: "This thing won't take off on me, will it?"

077 (1987)   The Ms. Harley-Davidson contest began in 1982 as the now forgotten "Harley Girl." The "goddess-like figure with upstretched arms that become wings" decorated derby and timing covers, gas cap medallions, belt buckles, and T-shirts with its quasi-ancient Egyptian or Valkyrie look. Maybe that scared riders away, because shortly after that the Harley Girl items were dropped and the contest became Ms. Harley-Davidson. This lasted until 1995 when it was quietly discontinued. Here Ms. H-D contestants are seen at Daytona in 1987.

078 (1988)   Harley-Davidson's 85th Anniversary ("the Homecoming") was the Motor Company's first big birthday bash since 1978. Milwaukee had plenty to celebrate after surviving the financial turmoil of the buy-back and having recaptured nearly 46 percent of the big-bike market. The Harley-Davidson Motor Co. was once again in the black and all stops were pulled as company execs repeated their group-ride theme by leading ten routes from around the continent that converged on Milwaukee's lakefront Summerfest grounds in June.

079 (1988)   What do you do for a birthday party? You bake a cake and decorate it. This over-sized example helped celebrate H-D's 85th Anniversary in Milwaukee — although it's doubtful there was enough to go around for the 35,000 enthusiasts who showed up for this bash.

080 (1988)   Although several H-D executives took a turn at the dunk tank as part of the Muscular Dystrophy Association fund-raiser during the 85th Anniversary celebration, none was more popular than Ms. Harley-Davidson for 1988 — Jackie McCue. The Homecoming raised more than a half-million dollars and boosted Harley's total fund-raising goal for MDA that year to $1.5 million.

081 (1988) Today, the term "hog" is universally accepted as a nickname for the Harley-Davidson Big Twin. One logical theory of the term's origin points to a 1920 photo of H-D team racer Ray Weishaar and his "piggy mascot." But there was also a Harley-riding publicist back then named Johnny Hogg. Some modern riders scoff at these romantic explanations and claim that "hog" originated with English bike riders in the 1950s who looked down their noses at Harley full dressers and called them "beasts" or "road hogs," later shortened simply to "hog."

082 (1988) By 1988 the Low Rider looked a lot different from the 1977 original. In 1986 the Low Rider name jumped from the discontinued 4-speed FXSB to the 5-speed FXRS Low Glide. The FXR's rubber-mounted chassis pushed Harley styling into a new direction. With plastic side covers, angular frame tubes, and hidden battery/oil tank, the FXR made some Harley traditionalists uneasy. Yet this bike was perfect for the ex-import rider, who wanted a smooth and better-handling Harley. The Japanese had been trying to build a bike like this for years. Harley-Davidson made it look easy.

083 (1988) The Springer Softail was Milwaukee's most radical factory custom to date when it debuted in 1988. Who else but Harley-Davidson would dare bring back the old springer fork? Trouble was, the springer fork had been retired for thirty years and nobody around the factory remembered how to build it. So Harley engineers went to visit John Nowak, who knew just about everything about the old factory. They sat on the floor of Nowak's den while "Mr. Motorcycle" explained the ins and outs of the springer manufacturing process. It was the last act in a career that went back to 1936.

084 (1989) When the Fat Boy appeared in 1990, Harley-Davidson was coming back strong and ready to rumble in the heavyweight motorcycle arena. Still, the Fat Boy took everyone by surprise. Willie G. Davidson had infused powerful nostalgic elements into this premier custom Softail offering. Enthusiasts worldwide took note of this model's special features: solid disc wheels, a monochromatic paint scheme in silver, both frame and sheet metal, yellow highlights on the engine and trim, shotgun exhausts, and a cut-back front fender. With the Fat Boy a burly new member of the family was born.

085 (1989) Then a professor at the University of Wisconsin, Martin Jack Rosenblum was focusing his creative energy on the "Ranger Project." This resulted in a volume of Harley-Davidson-inspired poetry (*The Holy Ranger*, 1989) and two musical CDs (*Free Hand*, 1991, and *Down on the Spirit Farm*, 1994). Described by the *Chicago Tribune* as "the Bob Dylan of the Harley-Davidson biker cult," this Ph.D.–wielding Harley rider today combines scholarship with poetic insight as Harley-Davidson's official historian.

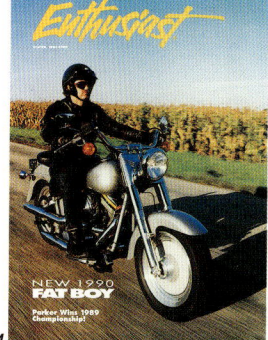

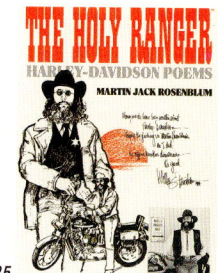

086 (1990) When the Fat Boy debuted in 1990, some riders believed its name was derived from the "Fat Man" and "Little Boy" atomic bombs of World War II. Its true origin, however, stems from the fat tires, disc wheels, and full fenders, which give it a bulky appearance. Urban myth debunked.

087 (1993) What at first glance might be taken for a Wide Glide is in fact a vanishing breed: an authentic chopper. Seen at Harley-Davidson's 90th Anniversary, this highly modified bike based on an Electra Glide exemplifies what guys regularly did to big full dressers years ago, and a faithful few still do. But with the wide variety of well-executed factory customs available today, it's hardly necessary. From the grime-covered pipes and sagging saddlebags, this chopper looks like it came the distance.

088 (1993) Custom paint jobs have long been a favorite means of personalizing a Harley. This flaming dragon-emblazoned example is a long way from the one-option-only black, gray, then green H-Ds of the early decades. Today, paint schemes range from stock, Genuine Harley-Davidson Parts & Accessories paint options, and to do-it-yourself spray bomb jobs, to professional creations like this one that can cost thousands of dollars.

089 (1993) Harley riders have always had a penchant for badges, patches, and pins to adorn their biker attire. Even in the 1930s, H-D was offering various emblems for enthusiasts. Because of the influence of motorcycle gangs and B-movies, for a time these ornaments took on a provocative and even offensive nature. The collection seen on this vest during H-D's 90th Anniversary, however, is representative of today's Harley rider.

090 (1993) It's not exactly clear just what the owner of this highly polished and whitewall-equipped Heritage Softail is trying to say here. As a security measure these handcuffs don't compare to the case-hardened, rubber-coated chains and disc locks of today.

091 (1993) Other manufacturers have tried to imitate the Harley Owners Group but without much success. That special bond of camaraderie among Harley riders doesn't seem to hold true for most other makes. The success of HOG can be seen in the numbers. From 33,000 members at the beginning of 1984, membership took off like a rocket to over 500,000 worldwide today and shows no sign of slowing down. One only wonders what HOG membership will be by the time of Harley-Davidson's 100-year anniversary in 2003 — a million?

092 (1994)   Back in the black financially, Harley-Davidson re-entered professional road racing with the VR1000. But in this case the only "chip off the old block" was the Bar & Shield logo on the fuel tank. While a V-Twin, the new VR was liquid-cooled, ran chain-driven double overhead camshafts, and had a 60-degree angle between the cylinders. While early victories were elusive, by 1999 the VR1000 was finally experiencing podium finishes in a highly competitive class.

093 (1995)   All rise and hail the king — King of the Highway, that is. It's pretty safe to say that as long as there is a Harley-Davidson Motor Company there will be a full-dress touring rig to carry that name. In 1995, H-D celebrated the Electra Glide's 30th birthday by issuing a special edition. That year also saw the introduction of fuel injection. Providing more power and improved performance, electronic fuel injection is now available on several H-D touring models.

094 (1997)   What do Cheeseheads do on a Sunday afternoon in January when the Packers are in the playoffs and the game's over? They go for a sidecar ride! That's me on the Road King with my friends Jean Fisher and Andrew Kirov in the tub. Their smiling faces demonstrate that the sidecar is a good-time vehicle in any season. If more enthusiasts tried winter riding, the sidecar's past popularity would surely return. Luckily, my old winter windscreen lined up with the Road King windshield's screws, almost as if some guardian angel at Harley-Davidson had planned it that way.

095 (1997)   Probably the most glamorous and widely anticipated Harley to come along in years was the Heritage Springer. If a bike ever exhaled pure class it's the FLSTS. It's also another unique factory restoration. After all, the first Springer had custom styling and strong chopper influence. Thus, when the Heritage Springer was prototyped at the factory, Willie G. and his stylists were almost like antique bike experts putting a chopper back to stock. By adding an older-style front fender and 16-inch wheel and tire, a new machine with a classic 1940s look was launched.

096 (1998)   Wish this were your garage? It's part of H-D's motorcycle collection at the Juneau Avenue Archives. For decades Harley has saved an example from each year's production. Today the collection numbers about 255 bikes and gaps in the collection continue to be filled in. Recent acquisitions include a 1925 Big Twin with double-wide sidecar, a 1934 Single, a replica *Easy Rider* chopper, and a 1987 Buell RR100. Plans for the new Harley-Davidson Museum in Milwaukee (to open in 2002) would put some of this fantastic collection and other memorabilia on permanent public display.

097

098

099

100

097 (1998)   This sea of bikes was seen at the 15th Anniversary HOG Rally at Milkaukee's State Fair Park in June of 1998. Three days of activities included music by several well-known bands, parades, races, a rodeo, and a two-day bike show. A huge wooden cake was pushed onto the stage at one point, and to the delight of onlookers the cake opened up and Willie G. and Nancy Davidson rode out on a 95th Anniversary Harley-Davidson motorcycle. Heavy rains during part of the rally could not dampen the enthusiasm of the 53,715 spirited HOG members that came from all 50 states and 42 other countries.

098 (1998)   Harley-Davidson anniversary celebrations just keep getting bigger and better. Even heavy rain that chased some riders across the country towards the 95th reunion did not dampen their spirits. On Saturday — the day of the big parade through Milwaukee and party at the lakefront — the heavens cleared and the faithful were blessed with clear blue skies. Plans that called for 25,000 motorcycles in the parade were scrapped when the bikes just kept coming. Over 50,000 Harleys rode through Milwaukee streets lined with cheering spectators wishing Harley-Davidson a happy 95th birthday.

099 (1999)   In 1999 an early Harley-Davidson engine was discovered in LaCrosse, Wisconsin. Acquired shortly afterwards by the Motor Company Archives in Milwaukee, this little "pee-wee" motor is the only known surviving example of a Harley-Davidson utility or "buckboard" engine. Tentatively dated to early 1905, this one still has its original angle-iron stand and was probably used to power a farm implement — possibly a cream separator. Here Willie G. Davidson inspects the little single in the Archives shortly after its acquisition with archives director Thomas C. Bolfert and H-D historian Martin Jack Rosenblum.

100 (2000)   Motorcycles are a romance of moon-lit rambling across the dreamscape of a soul. No bike better demonstrates the power of this experience than Night Train. With flowing steel lines inherited from the 1936 Knucklehead, the cocky persona of a 1960s performance chopper, and the technological expertise of the Twin Cam 88B counter-balanced engine, Night Train speaks a deep and primal truth. A language that Harley-Davidson first uttered in 1903 with their black, hand-brushed single conceived in part in the Milwaukee railshops where old Bill Davidson swung a hammer, and that applies today to this classically inspired and perfectly executed dream baby.

This book was printed in four colors on a sheetfed printing press on 157 gsm Japanese matte art paper. The case has been covered in T-Saifu linen cloth. The medal plate adhered to the debossed area on the front case of the book is made of brushed aluminum that has been silkscreened in two colors. The font used in the book is Trade Gothic.